Sea stars live

in the sea.

Superstars of the Sea

SEA STARS

Theresa Emminizer

PowerKiDS press

PK Beginners

Let's look at sea stars!

5

Sea stars are not fish!

Sea stars are soft.

9

This sea star is red.

11

This sea star is purple.

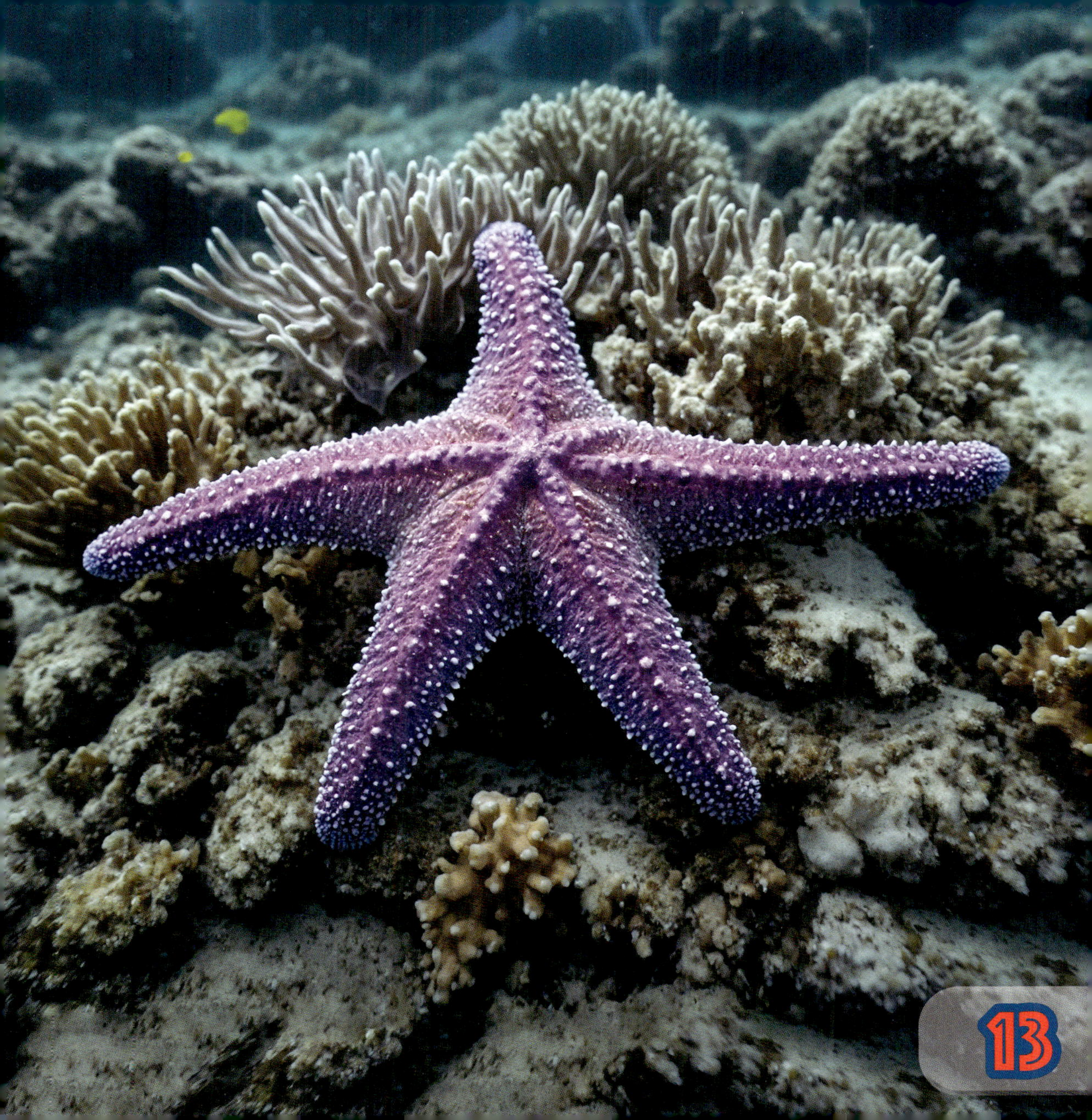
13

This sea star is orange.

This sea star is blue.

17

This sea star has
five arms.

19

This sea star has many arms!

21

Sea stars are special.

23

Published in 2026 by The Rosen Publishing Group, Inc.
2544 Clinton Street, Buffalo, NY 14224

First Edition

Editor: Theresa Emminizer
Book Design: Jeffrey Taylor

Photo Credits: Cover Vojce/Shutterstock.com; p. 3 Jesus Cobaleda/Shutterstock.com; p. 5 FamVeld/Shutterstock.com; p. 7 Damsea/Shutterstock.com; p. 9 Paulo Violas/Shutterstock.com; p. 11 Cigdem Cooper/Shutterstock.com; p. 13 Shahidphotographer/Shutterstock.com; p. 15 Vojce/Shutterstock.com; p. 17 blue-sea.cz/Shutterstock.com; p. 19 Rob Atherton/Shutterstock.com; p. 21 KONSTANTIN_SHISHKIN/Shutterstock.com; p. 23 Damsea/Shutterstock.com.

Cataloging-in-Publication Data

Names: Emminizer, Theresa.
Title: Sea stars / Theresa Emminizer.
Description: Buffalo, New York : PowerKids Press, 2026. | Series: Superstars of the sea identifiers: ISBN 9781499451559 (pbk.) | ISBN 9781499451566 (library bound) | ISBN 9781499451573 (ebook)
Subjects: LCSH: Starfishes--Juvenile literature.
Classification: LCC QL384.A8 E46 2026 | DDC 593.9'3--dc23

Manufactured in the United States of America

Some of the images in this book illustrate individuals who are models. The depictions do not imply actual situations or events.

CPSIA Compliance Information: Batch #CSPK26. For further information contact Rosen Publishing at 1-800-237-9932.